FUN FACTS!

DID YOU KNOW?

By Tatyana L. Walton

DID YOU KNOW?

scientific fun facts

By Tatyana L. Walton

Did you know it snows in the Sahara Desert?

SNOW IS VERY RARE IN THE DESERT BECAUSE THERE IS NOT ENOUGH WATER IN THE AIR FOR IT ,EVEN THOUGH IT CAN GET VERY COLD AT NIGHT.

DID YOU KNOW COLOMBIA'S BRIGHTEST RAINBOW IS IN ITS RIVER?

IT'S CALLED CANO CRISTALES

IT CHANGES COLOR BECAUCE OF THE MACARENIA CLAVIGERA PLANT

THE COLORS DEPEND ON LIGHT AND WATER CONDITIONS

THE PLANT HAS BEEN SEEN IN GREEN, ORANGE, RED, YELLOW, AND EVEN GREEN

THE RIVER HAS BEAUTIFUL CIRCULAR ROCKS POOLS

DID YOU KNOW A HIPPO'S JAW OPENS WIDE ENOUGH TO FIT A SPORTS CAR INSIDE?

HIPPOS LOVE WATER AND THEY SPEND MOST OF THE DAY IN IT TO STAY COOL. THE HIPPO CAN EVEN BREATHE, SEE, AND HEAR WHILE ITS BODY IS UNDER WATER BECAUSE ITS NOSE, EARS, AND EYES ARE ON THE TOP OF ITS HEAD. DO HIPPOS SWIM BETTER THAN PEOPLE? YES, THEY ARE EXCELLENT SWIMMERS AND CAN HOLD THEIR BREATH FOR FIVE MINUTES.

DID YOU KNOW IT WOULD TAKE 19 MINUTES TO FALL FROM THE NORTH POLE TO EARTH'S CORE?

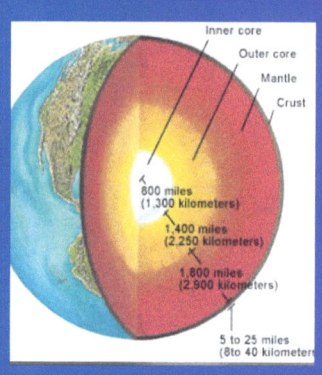

THE CORE IS ALMOST THE SIZE OF THE MOON. THE EARTH'S INNER CORE IS SURPRISINGLY LARGE, MEASURING 2,440 KM (1,516 MILES) ACROSS. ...
IT'S HOT...REALLY HOT. ...
IT'S MOSTLY MADE OF IRON.
...
IT SPINS FASTER THAN THE SURFACE OF THE EARTH. ...
IT CREATES A MAGNETIC FIELD.

DID YOU KNOW WATER MAKES DIFFERENT SOUNDS DEPENDING ON ITS TEMPERATURE?

TEMPERATURE IS ANOTHER CONDITION THAT AFFECTS THE SPEED OF SOUND. HEAT, LIKE SOUND, IS A FORM OF KINETIC ENERGY. MOLECULES AT HIGHER TEMPERATURES HAVE MORE ENERGY AND CAN VIBRATE FASTER AND ALLOW SOUND WAVES TO TRAVEL MORE QUICKLY. THE SPEED OF SOUND AT ROOM TEMPERATURE AIR IS 346 METERS PER SECOND.

DID YOU KNOW SNAILS CAN SLEEP UP THE 3 YEARS ?

YES! IF THE WEATHER DOESN'T MEET THEIR NEEDS

SNAILS WILL SLEEP ON AND OFF FOR SEVERAL HOURS AT A TIME.
ONCE THEY'VE RESTED THOUGH, THEY CAN STAY AWAKE FOR AROUND 30 HOURS !

SLUGS AND SNAILS HIDE IN DAMP PLACES DURING THE DAY. THEY STAY UNDER LOGS AND STONES OR UNDER GROUND COVER. THEY ALSO HIDE UNDER PLANTERS AND LOW DECKS. AT NIGHT THEY COME OUT TO EAT

DID YOU KNOW HONEYBEES FLAP THEIR WINGS 230 TIMES EVERY SECOND?

THEIR WINGS BEAT OVER A SHORT ARC OF ABOUT 90 DEGREES, BUT RIDICULOUSLY FAST, AT AROUND 230 BEATS PER SECOND

BEES FLAP THEIR WINGS WHILE REMAINING STATIONARY TO MOVE AIR THROUGHOUT THE BEEHIVE TO REGULATE TEMPERATURE, TO SPREAD PHEROMONES THROUGHOUT THE HIVE, AND TO EVAPORATE NECTAR MOISTURE. INSECTS LIKE HONEY BEES THAT HAVE FOUR WINGS ARE CALLED HYMENOPTERANS WHILE TWO-WINGED INSECTS LIKE MOSQUITOES ARE CALLED DIPTERANS.

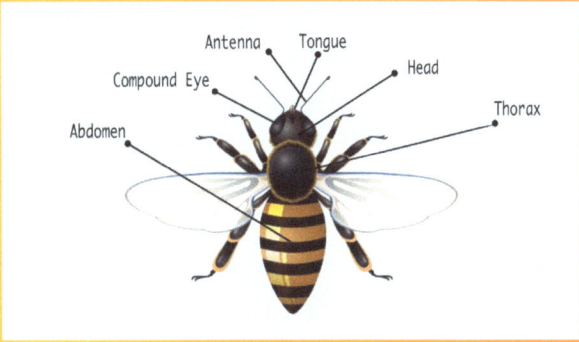

DID YOU KNOW BROCCOLI CONTAINS MORE PROTEIN THAN STEAK?

RIBEYE STEAK HAS OVER 29 GRAMS OF PROTEIN PER 3.5 OUNCES, WHEREAS PRIME RIB STEAK BOASTS 22.5 GRAMS. BY COMPARISON, A 3.5-OUNCE SERVING OF COOKED BROCCOLI HAS ROUGHLY 2.4 GRAMS OF PROTEIN.

DID YOU KNOW M&MS ARE NAMED AFTER THEIR CREATORS: MARS & MURRIE?

WHEN M&M'S FIRST HIT THE MARKET IN 1941, THE ORIGINAL COLORS WERE RED, YELLOW, GREEN, BROWN, AND, GUESS WHAT? PURPLE. THIS VARIETY OF COATED CANDIES WAS SENT AROUND THE WORLD DURING WORLD WAR II IN ITS ORIGINAL CARDBOARD TUBE PACKAGING

DID YOU KNOW WATER CAN EXIST IN THREE STATES AT THE SAME TIME?

THIS IS CALLED THE TRIPLE BOIL, AND AT THAT TEMPERATURE, WATER EXISTS AS A GAS, A LIQUID, AND A SOLID SIMULTANEOUSLY.

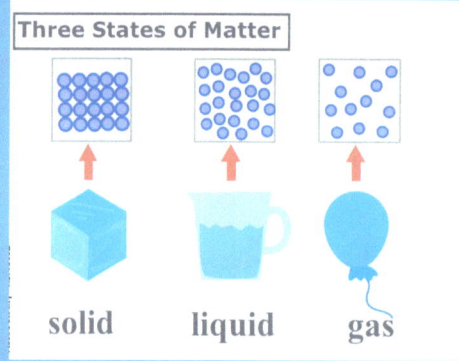

DID YOU KNOW THE MAJORITY OF EARTH'S OXYGEN IS PRODUCED BY OCEANS?

ROUGHLY HALF OF THE OXYGEN PRODUCTION ON EARTH COMES FROM THE OCEAN. THE MAJORITY OF THIS PRODUCTION IS FROM OCEANIC PLANKTON — DRIFTING PLANTS, ALGAE, AND SOME BACTERIA THAT CAN PHOTOSYNTHESIZE. ONE PARTICULAR SPECIES, PROCHLOROCOCCUS, IS THE SMALLEST PHOTOSYNTHETIC ORGANISM ON EARTH

DID YOU KNOW THE HUMAN BRAIN IS 78% WATER?

THE BRAIN AND HEART ARE COMPOSED OF 73% WATER, AND THE LUNGS ARE ABOUT 83% WATER. THE SKIN CONTAINS 64% WATER, MUSCLES AND KIDNEYS ARE 79%, AND EVEN THE BONES ARE WATERY: 31%

DID YOU KNOW A CROCODILE CANNOT STICK ITS TONGUE OUT?

CROCODILES HAVE A MEMBRANE THAT HOLDS THEIR TONGUE IN PLACE ON THE ROOF OF THEIR MOUTH SO IT DOESN'T MOVE. THIS MAKES IT IMPOSSIBLE FOR THEM TO STICK IT OUTSIDE OF THEIR NARROW MOUTHS. THAT CAN BE HANDY FOR THE REPTILE WHEN SNAPPING ITS JAWS SHUT RAPIDLY.

DID YOU KNOW A SHRIMP'S HEART IS IN ITS HEAD?

THE HEART IS LOCATED ON THE THORAX RIGHT PASSED ITS HEAD. HOWEVER, A SHRIMP IS COVERED BY A SINGLE EXOSKELETON, SO THEREFORE, SOME MISTAKE THE THORAX FOR A PART OF ITS HEAD

DID YOU KNOW YOU TYPICALLY ONLY BREATHE OUT OF ONE NOSTRIL AT A TIME?

ONE NOSTRIL WORKS AT A TIME WHILE WE BREATHE IN AND OUT THROUGHOUT THE DAY,
AND THEY SWITCH EVERY FEW HOURS.
THIS IS CALLED THE NASAL CYCLE,
A PROCESS OF ALTERNATING CONGESTION AND DECONGESTION

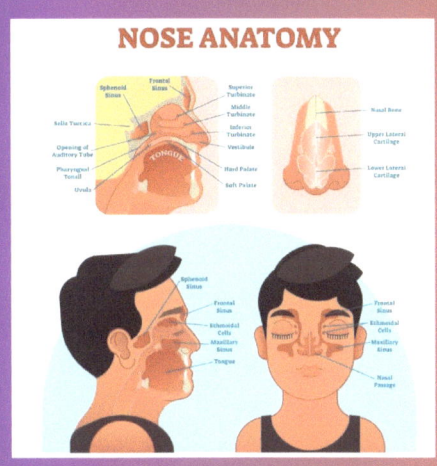

DID YOU KNOW STOP SIGNS USED TO BE YELLOW?

YELLOW WAS THE STANDARD COLOR FOR STOP SIGNS FOR NEARLY 30 YEARS. THE FIRST STOP SIGN APPEARED IN 1915 IN DETROIT, MICHIGAN. THERE WERE A VARIETY OF COLORS USED FOR STOP SIGNS UNTIL THE LATE 1920S, WHEN THE BACKGROUND COLOR WAS STANDARDIZED ON YELLOW FOR MAXIMUM DAY AND NIGHT VISIBILITY

From 1924 to 1954, stop signs were yellow octagons with black letters.

Did you know The fastest gust of wind ever recorded on Earth was 253 miles per hour?

Barrow Island, Australia • April 10, 1996
Wind trace taken at Barrow Island, Australia, during Tropical Cyclone Olivia. The wind scale is in meters per second. The peak gust of 113.2 m/s (253 mph) occurred at around 6:15 pm local time.

DID YOU KNOW A LION'S ROAR CAN BE HEARD FROM FIVE MILES AWAY?

It's so loud it can reach 114 decibels (at a distance of around one metre) and can be heard from as far away as five miles. This volume is all to do with the shape of the cat's larynx

Did you know The nearest star to Earth is 4.2 light-years away?

The closest star, Proxima Centauri, is 4.24 light-years away. A light-year is 9.44 trillion km, or 5.88 trillion miles. That is an incredibly large distance. Walking to Proxima Centauri would take 950 million years

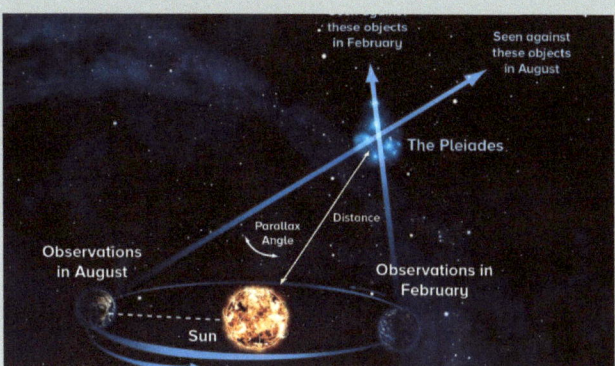

Did you know Sudan has the most pyramids in the world?

There are around 2000 Kushite pyramids in upper Sudan, compared with 200 Egyptian pyramids. Showing the relationship between the African civilizations, the Kushite pyramids depict bilateral trade, movement of people and knowledge.

Did you know 25 is the sum of the five consecutive single-digit odd natural numbers?

25 is a centered octagonal number, a centered square number, a centered octahedral number, and an automorphic number. 25 percent (%) is equal to ¼.

DID YOU KNOW?

scientific fun facts

By Tatyana L. Walton

www.ingramcontent.com/pod-product-compliance
Lightning Source LLC
Chambersburg PA
CBHW040058250526
45473CB00043B/1876